ARTIFICIAL INTELLIGENCE IN ACTION

SELECTED ADB INITIATIVES IN ASIA AND THE PACIFIC

MAY 2024

ADB

ASIAN DEVELOPMENT BANK

© 2024 Asian Development Bank
6 ADB Avenue, Mandaluyong City, 1550 Metro Manila, Philippines
Tel +63 2 8632 4444; Fax +63 2 8636 2444
www.adb.org

Some rights reserved. Published in 2024.

ISBN 978-92-9270-687-6 (print); 978-92-9270-688-3 (PDF); 978-92-9270-689-0 (ebook)
Publication Stock No. SPR240246-2
DOI: http://dx.doi.org/10.22617/SPR240246-2

Notes:
In this publication, "$" refers to United States dollars.

Cover image by Jamillah Knowles & Reset.Tech Australia / Better Images of AI / Social media content / CC-BY 4.0.
Cover design prepared by Mike Cortes.

Contents

Foreword

Strategy 2030 from the Asian Development Bank (ADB) explicitly states that digital transformation is necessary to make the organization stronger, better, and faster. Established in 2018, the ADB Digital Innovation Sandbox is one of the programs that emerged to attain this ambitious goal. It serves as a safe and neutral facility for ADB to test and review new and emerging digital technologies in five areas—artificial intelligence (AI), robotics, distributed ledger technology, big data, and mixed reality—to see whether these have the potential to meet business and operations needs and drive the Digital Agenda 2030.

AI was still an emerging technology when the Digital Innovation Sandbox was established in 2018. Nonetheless, ADB saw its value and potential in helping deliver targets. Thus, the earliest initiatives that were tested through the Digital Innovation Sandbox were AI-driven. At that time, the work focused on machine learning. One such initiative was Robo-Advisor, which could answer questions related to ADB's operations related to gender, finance, and climate change, 6 years before the launch of ChatGPT. The emergence of deep learning opened even more opportunities for ADB to test AI in its work, leading to initiatives that used various data sources to predict economic trends. As of this time of writing, ADB is exploring generative AI, which is projected to introduce a new development era.

AI holds a lot of potential to change the world, but these changes can go both ways. The AI Governance Working Group, which comprises the Budget, Personnel, and Management Systems Department; Climate Change and Sustainable Development Department; Department of Communications and Knowledge Management; Information Technology Department; Office of the Auditor General; Office of the General Counsel; Office of the President; Office of Professional Ethics and Conduct; Office of Risk Management; Procurement, Portfolio, and Financial Management Department; Strategy, Policy, and Partnerships Department; Sectors Group; Southeast Asia Department; Office of the Vice-President (Administration and Corporate Management); and the operations departments, provides guidance on the ethical and responsible use of AI across ADB. By harnessing the technology, tempered by guiding principles such as fairness, inclusiveness, transparency, fair use, accuracy, and security, ADB can improve the effectiveness and efficiency of its operations, including the support it provides to developing member countries (DMCs).

With the rapid evolution of AI and the transition from the Digital Innovation Sandbox to the Digital Learning Labs, ADB is primed to discover new and innovative ways to use this digital technology, not only for future-proofing but also to help DMCs adapt to modern challenges. For example, AI-powered tools for climate modeling can help governments identify the anticipated impacts of climate change and take proactive measures. Similarly, AI-powered tools can be used to optimize food supply chains or predict disease outbreaks.

Before we look into the future, we must learn from our past experiences to help us grow and adapt. This publication provides a glimpse of some of the AI work ADB has done in Asia and the Pacific. It is worth noting that the bank did not just explore ways in which AI could be used to improve internal processes; rather, it also tested how the digital technology could be used to address real-world challenges in the areas of health, economy, transport, and social protection faced by DMCs. Most of these were implemented through the Digital Innovation Sandbox.

However, three tools—the Code of Conduct Chatbot, the Chatbots as Lifelines for Domestic Violence Survivors, and the Automatic Identification System to Generate Insights on Maritime Activities—were spearheaded by other ADB departments, an indication of the growing role of AI in the bank's work.

I hope that this publication will spur further research and development on ways in which AI and other digital technologies could be leveraged to foster inclusive and prosperous development for the benefit of all.

Muhammad Ehsan Khan
Director General, East Asia Department and
Intelligence Governance Working Group Chair
Asian Development Bank

Acknowledgments

Artificial Intelligence in Action: Selected ADB Initiatives in Asia and the Pacific was developed by the Information Technology Department (ITD) of the Asian Development Bank (ADB).

Marc Lepage, principal information technology specialist for Technology Innovation at ITD, conceptualized and guided the development of this knowledge product. Ozzeir Khan, ITD director, provided guidance.

Melanie Sison, technical writer, prepared the case studies with support from the following ADB colleagues, who provided technical input and feedback on the documentation of initiatives that involved their respective offices:

- Pierre Dyens, principal human resource specialist at the Budget, Personnel, and Management Systems Department;
- Veronica Joffre, senior gender and social
- development specialist at the Climate Change and Sustainable Development Department;
- Nelly Defo, senior financial control
- specialist at the Controller's Department;
- Pinky Serafica, senior communications officer at the Department of Communications and Knowledge Management;
- Abdul Abiad, director of the Economic Research and Development Impact Department (ERDI);
- Elaine S. Tan, advisor at ERDI and head of the Statistics and Data Innovation Unit;
- Homer Pagkalinawan, associate economics officer at ERDI;
- Madhavi Pundit, senior economist at ERDI;
- Maya Vijayaraghavan, principal evaluation specialist at the Impact Evaluation Department;
- Kadra Saeed, principal information technology specialist (Digital Products Lead) at ITD;
- Christel Adamou, director of the Office of Professional Ethics and Conduct;
- Maria Lorena Cleto, social development specialist (Resettlement) at the Office of Safeguards;
- Fiona Alpe, principal transaction support specialist at the Private Sector Operations Department;
- Xiufeng Zhao, principal portfolio management specialist at the Procurement, Portfolio, and Financial Management Department;
- Alaysa Tagumpay Escandor, public management officer (Governance) at Sectors Group (SG);
- Arun Ramamurthy, principal transport specialist at SG;
- Pawan Karki, principal transport specialist at SG; and
- Jae Kyoun Kim, health specialist at SG.

Special thanks go to ITD director general Stephanie Hung and East Asia Department director general and Artificial Intelligence (AI) Governance Working Group chair Muhammad Ehsan Khan for their invaluable support in the development of this publication.

The team thanks its partners, the Asian Institute of Management and Oracle Labs. In addition, acknowledgment is given to technology service providers that worked with ADB to develop the AI-powered initiatives featured in this knowledge product: ORIS, Neural Mechanics Inc., and Yellow.ai.

Credit for the images in this publication goes to Better Images of AI, a nonprofit collaboration of various individuals and nonprofit and academic institutions that advocates for the correct representation of AI, considering that many publicly available images often misrepresent the technology and reinforce negative or even harmful stereotypes.

Abbreviations

ADB	Asian Development Bank
AI	artificial intelligence
AIS	Automatic Identification System
AOI	area of interest
COVID-19	coronavirus disease
CTLA-TA	Controller's Department Loan Administration Division—Technical Assistance Section
DBSCAN	Density-Based Spatial Clustering of Applications with Noise
DMC	developing member country
EIM	economic indicator model
EOI	event of interest
GDP	gross domestic product
GPS	global positioning system
H3	Hexagonal Hierarchical Geospatial Indexing System
ICP	Intelligent Concept Paper
IDD	integrity due diligence
IED	Independent Evaluation Department
IT	information technology
ITD	Information Technology Department
km	kilometer
LAR	land acquisition and resettlement
MARI	MyADB Recruitment Intelligence
MVP	minimum viable product
NCAV	National Center Against Violence (Mongolia)
NLI	National Legal Institute (Mongolia)
OPEC	Office of Professional Ethics and Conduct
PAI	Project Administration Instructions
PIA	PPFD Intelligent Assistant
POC	proof of concept
PPFD	Procurement, Portfolio and Financial Management Department
PSOD	Private Sector Operations Department
RT-PCR	reverse transcription polymerase chain reaction
SDG	Sustainable Development Goal
TA	technical assistance
TAM	theme augmented model
TAMI	Technical Assistance Messaging Intelligence
TASU	Technical Assistance Supervising Units

Introduction from the Information Technology Department Director General

The developing member countries (DMCs) of the Asian Development Bank (ADB) face increasingly challenging circumstances. Even as they pivot toward recovery after being heavily hit by the coronavirus disease (COVID-19), they also have to contend with climate change, biodiversity loss, and other issues that threaten not only to hinder, but reverse the hard-earned development progress that they have striven for in the last decades.

In the past years, ADB's role has evolved from being a financing partner to also serving as a knowledge bank that enables the development of DMCs. Knowledge is the ground on which innovation stands, and innovation, in turn, can accelerate economic growth and social progress, especially in the context of knowledge as a public good. This also implies that putting minimal focus on knowledge and innovations can put not just the bank, but also the DMCs it serves, at risk of being left behind.

Artificial intelligence (AI), particularly generative AI, while not a silver bullet, can be a powerful tool that would help put us on the right path to address these growing challenges. It has the potential to serve as a key driver of growth and innovation. However, this can only be fully realized if sufficient investments are made not only to effectively harness the technology but also to empower teams to become competent in responsibly leveraging AI to augment decisions and enhance productivity and efficiency. After all, ADB's digital transformation does not just refer to the hardware and software; the capacity of its personnel dictates the speed at which technology will be adopted by the bank.

At the Information Technology Department, we are exploring ways in which AI can be leveraged to redefine the way we work, and, more importantly, craft solutions that would help address the increasingly multifaceted issues faced by Asia and the Pacific, particularly the poorest and most vulnerable communities in the region. We are not only collaborating with other departments in ADB and with stakeholders in DMCs; we are also pursuing and nurturing partnerships with industry leaders and innovative startups to achieve our goal of responsible AI development, allowing us early access to test new tools and technologies to see whether they are compatible with ADB's systems and processes. Collaboration also offers additional benefits in the context of AI innovations: by leveraging the collective intelligence and experience of experts from various fields and backgrounds, it can accelerate progress. It can also help mitigate biases and ethical issues in AI development.

The work we have done in AI in the past years has resulted in the accumulation of knowledge, and this publication provides us with snapshots of some of what we have learned in the last 6 years. We hope that you can view these use cases through a growth-mindset lens. AI is still in the nascent stages of development, which makes this the best time for us to test, experiment with, and learn different ways we can use AI in different contexts. Let us learn and grow together.

Stephanie King-chung Hung
Chief Information Officer and
Director General, Information Technology Department
Asian Development Bank

CHAPTER 1

Artificial Intelligence in Development Operations

This image depicts the complexity of AI: what may be perceived as a single entity is actually several layers of algorithms linked together to form a cohesive system.

Image by Anton Grabolle / Better Images of AI / AI Architecture / CC-BY 4.0

As a development, knowledge, and climate bank, the Asian Development Bank (ADB) strives to support developing member countries (DMCs) in Asia and the Pacific by offering knowledge, expertise, and financing. The region faces increasingly complex challenges such as climate change, poverty, and food insecurity. Addressing these issues requires forward thinking and new solutions.

ADB is exploring how modern technologies such as artificial intelligence (AI) could be used in its operations. This section shows different use cases of AI at various stages of the project cycle, from the time the concept paper of a project is written to extracting lessons from project evaluations. It also features ways in which the technology was used in the areas of road design, health, transport, and social protection. In the future, new use cases of AI will likely emerge as the world discovers different ways in which it can be used to resolve complex issues.

Intelligent Concept Paper—Problem Tree Generator

Writing a concept paper is the first step ADB takes to support DMCs. To do so, writers need to manually sift through various internal documents and online materials; as a result, a single concept paper can take between several weeks and a month to prepare. Problem trees, which show the causes and effects of an issue that ADB seeks to help resolve, can alone take 3 to 5 days to complete. Even then, concept papers are unlikely to be comprehensive given limitations in finding available references. Thus, ADB sought to use technology to make writing concept papers faster and more efficient. This initiative explored using AI to automate the generation and population of problem trees based on entered search queries.

The Information Technology Department (ITD) engaged Neural Mechanics Inc. to develop a minimum viable product (MVP) of the tool that could be used to automatically generate problem trees. AI was leveraged to increase efficiency and reduce inconsistencies in preparing concept papers, particularly problem trees, because it was able to process large amounts of data. The tool, which would come to be known as Intelligent Concept Paper (ICP), was trained on similarity matching, enabling it to filter sentences that are semantically similar to texts from previous problem trees. A neural network model, which is a subset of machine learning that can identify relationships based on attributed weights, was then used to identify causal event pairs from the filtered sentences.

The sources of information were documents from ADB's East Asia Department and the World Bank, as well as Xinhua news articles. These were first converted to a readable format to enable the designed web scraper to crawl through these documents. These were annotated to make searching easier for the engine. The annotated materials were then stored in the cloud. An app was developed to recover relevant information based on the keywords entered by users.

SUMMARY

What it is:
The Intelligent Concept Paper was an artificial intelligence-powered system that automatically generated problem trees, shortening the time needed to prepare concept papers.

Start of implementation:
2019

Implementation partner:
East Asia Department

Technology service provider:
Neural Mechanics Inc.

In line with ADB's operational approaches:
- Promoting digital development and innovative technologies
- Applying differentiated approaches
- Delivering integrated solutions

Users entered their own problem statements or used/edited the suggestions that were auto-generated based on their entered keywords. From there, they had the option to use the suggested causes and effects or write their own. It was possible to add multiple responses for both causes and effects, as well as sub-causes and sub-effects, if users wanted to provide more specific information.

A click-and-drop feature was also incorporated into ICP's design to build the problem tree right on the platform. Users were able to share this with others for their review and input.

These features allowed ICP to generate results within seconds, significantly cutting down the time needed to research information and produce the problem tree diagram. The tool also allowed various users to collaborate on the same problem tree and document, making work more efficient.

Work on ICP began in 2019 and was completed in April 2020. While this tool is no longer in use, the initiative showed that it was possible to develop an AI-driven platform that built problem trees. The tool marked various firsts for ADB: the first homegrown AI for ADB operations, the first one developed that leveraged the wealth of knowledge from previous works of ADB within the development landscape, and the first to harness both technology and human inputs to scan through key materials to identify problems in ADB's DMCs.

ADB-wide consultations and/or focus group discussions may be conducted to explore how lessons from the tool development can be used in the future, including in writing other sections of a concept paper such as the Design and Monitoring Frameworks.

Digital Assistant for Project Administration Instructions

Project Administration Instructions (PAIs) are among the most accessed materials on the ADB website, with over 20,000 views and downloads yearly. The several series of PAIs provide information on the policies and procedures that should be followed throughout the entire project cycle of ADB-financed loan and technical assistance (TA) projects.[1]

Users spent from a few minutes up to several hours finding the information they needed. To make it easier to find key instructions and guidelines in PAIs, Yellow.ai, then called Yellow Messenger, was engaged to develop a digital assistant. The intent was to have a tool that could quickly provide users with the information they need without having to manually sift through various documents on the ADB website.

Work on the MVP took place between June and September 2020. Key personnel from the Procurement, Portfolio and Financial Management Department (PPFD) tested it between September and October 2020, and comments and feedback were considered in the development of the digital assistant, which was officially launched in October 2020. The digital assistant, which came to be known as PPFD Intelligent Assistant (PIA), continues to be used in 2024.

The virtual assistant is powered by AI, enabling it to interact with users in a conversational manner and generate results relevant to queries. Traditional keyword searches relied on users' familiarity with specific terms. In contrast, AI with natural language processing can be trained to generate results according to search intent. This allows PIA to learn with every new interaction, as well as from users' queries and feedback, enabling it to adapt to business needs.

PIA automatically generates links to documents, informative videos, and comprehensive user manuals in response to user queries. PIA was also designed to automatically pick up and furnish the latest available

SUMMARY

What it is:
The Digital Assistant for Project Administration Instructions makes finding information on project implementation easier for ADB clients and staff.

Start of implementation:
2020

Implementation partner:
Procurement, Portfolio and Financial Management Department

Technology service provider:
Yellow.ai

In line with ADB's operational approach:
- Promoting digital development and innovative technologies

information to users because PAI documents and technical guidance notes are updated on an ad hoc basis. Offline support from PPFD personnel is also available in case PIA is unable to provide the required information.

PIA also comes with an AI-powered analytical dashboard that is accessible to key ADB personnel who want to generate insights on the type of information clients and staff frequently search for.

The tool included special instructions related to the coronavirus disease (COVID-19) at the time the digital assistant was launched, which was when pandemic-related movement restrictions were in place. This became very handy, as clients and staff were able to get information in real time.

---

[1]   ADB. 2023. *Project Administration Instructions (PAI)*. Manila.

Around 1,200 users had over 1,700 sessions using PIA between October 2020 to June 2021. Each session was around 2 minutes long, which indicated that users read the information generated by the digital assistant.

The digital assistant was able to improve operational efficiency by enabling internal and external users to find information quickly and easily. At the same time, ADB used the dashboard feature to review common queries, feedback, and complaints from users, allowing it to introduce improvements. These elements contributed to fostering good relations between ADB and its clientele.

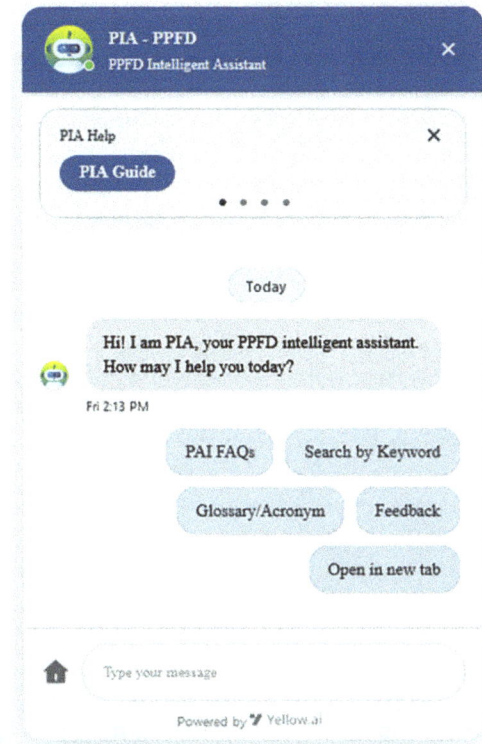

PIA on the Project Administration Instructions page on the ADB website.

Digital Twins for Smart Road Construction

ADB supported the rehabilitation of the Central Asia Regional Economic Cooperation Corridor in Uzbekistan, including a 25-kilometer (km) section of Section 1 of Highway A380, located in the Republic of Karakalpakstan and one of the key trade routes in the region. The Digital Twins for Smart Road Construction initiative aimed to improve not just connectivity but also road conditions and safety. Using a climate-resilient and sustainable design was also important, especially considering that Uzbekistan is experiencing water scarcity. Road construction projects consume large amounts of natural resources and contribute to significant global emissions. Cement manufacturing alone produces 5%–8% of global anthropogenic greenhouse gas emissions. About 22% of global carbon dioxide emissions come from activities related to roadwork, including mining, transportation, and paving, as well as emissions from road users.[2]

ADB's Climate Change Operational Framework 2017–2030 supports the shift toward low greenhouse gas emissions and climate-resilient development.[3] Therefore, a tool was needed that would support the shift toward environmentally friendly designs that provide value for money without compromising road safety and durability.[4]

To this end, ADB engaged the technology service provider ORIS, which offers a digital platform capable of running simulations to compare how different materials used in pavement solutions would perform in certain traffic conditions and different environments, including the potential impacts of climate change. The platform enables users to consider which road designs are environmentally friendly solutions that best fit the unique contexts where the connective infrastructure would be constructed.

The technical service provider, in collaboration with ADB and the Government of Uzbekistan, conducted field research and surveyed the quarries and material

SUMMARY

What it is:
The platform, which had a database of construction materials and was powered by artificial intelligence, allowed users to compare analyses and simulations of road designs and available local materials using multiple scenarios to determine which pavement solutions were more sustainable, resilient, cost-efficient, and best fit the local context.

Start of implementation:
2021

Implementation partners:
* Central and West Asia Department
* Government of Uzbekistan

Technology service provider:
ORIS

In line with ADB's operational priorities:
* Tackling climate change, building climate and disaster resilience, and enhancing environmental sustainability
  ‣ Mitigation of climate change increased
  ‣ Climate and disaster resilience built
  ‣ Environmental sustainability enhanced
* Promoting rural development and food security
  ‣ Rural development enhanced
* Fostering regional cooperation and integration
  ‣ Greater and higher quality connectivity between economies
* Strengthening governance and institutional capacity
  ‣ Strengthened country systems and standards

---

[2]   P. Newman et al. 2013. *Reducing the Environmental Impact of Road Construction*.

[3]   ADB. n.d. *ADB's Work on Climate Change and Disaster Risk Management*.

[4]   P. Karki. 2023. *Smart Ways to Make Uzbekistan's Road Networks Climate-resilient*. *Development Asia*. 5 April.

producers near the project site to get information on the available materials (including water), suppliers, and corresponding costs. AI was also used to gather data on geography, construction materials, weather forecasts, and pavement solutions; and the carbon impact of the assessed materials was analyzed. Uzbekistan regulations and international road design standards were also gathered.

Multi-criteria analyses and evaluations were then performed to determine the best design options. Among the criteria that were considered were the material types and costs, projected carbon emissions, and durability and resilience of the roads given the projected level of road use and impacts of climate change. The option to recycle good quality materials from the existing roads, such as asphalt, crushed stones, and shoulder materials, to further reduce costs was also considered. The ORIS platform was then used to perform a pavement design analysis to compare three designs (base case design, French, and German) to identify improvement opportunities to lower costs, spare natural resources, and support the recycling of materials. The platform was also used to generate recommendations on maintenance, climate change and sustainability optimization, and road safety improvement. Forty years' worth of data were also used to model how the road project is anticipated to experience the impacts of climate change, including heat, floods, and frost.

Two additional modules were integrated based on specific ADB requirements: a carbon dioxide calculator and data exchange on the road life cycle vehicle consumption. The old road safety standards were also updated, with over 10 normative documents and handbooks developed.

The improved design complied with United Nations (UN) global road safety performance targets for road construction. Analysis showed that using the improved road design reduced carbon emissions by 17% and natural resources use by 29%. Five million liters of water were saved, which is especially

important for a water-scarce country such as Uzbekistan. Costs went down by 10%, which was equivalent to $3.2 million. The analysis also showed that a $7.1-million investment in the project would make it exceed the minimum UN 3-star safety target, resulting in a 56% reduction of fatalities along that section.

These results were achieved while delivering the other benefits of road rehabilitation, including reducing travel time by 15% and supporting trade in the region.

The Digital Twins for Smart Road Construction initiative had a second phase, which covered the rehabilitation of a 4-km-long section of the 4R180 rural road in Uzbekistan and also incorporated climate-resilient design elements. This was the first performance-based maintenance contract in the country.

This initiative has shown that the digital twin platform can be used as a tool for due diligence.[5] Its ability to conduct multi-criteria analyses can aid decision-makers in project implementation, including developing sustainable road designs. This is important given the growing challenges due to climate change. Other types of technology may also be considered in similar projects in the future. For example, drones and sensors may be used separately or alongside digital twins to further improve the monitoring of road conditions and durability.

---

[5] ADB. 2023. *Building Greener, Resilient Transport Infrastructure: Innovating with Artificial Intelligence and Digital Twins in Road Design in Uzbekistan*. Manila.

Chatbots as Lifelines for Domestic Violence Survivors

COVID-19 triggered a viral pandemic, but its impact went beyond health: data also showed a surge in domestic violence cases against women and girls since the pandemic started.[6] Mongolia, where approximately one in three women have experienced domestic violence at some point in their lives, was not spared from this troubling trend.[7] Police reports from Ulaanbaatar alone reflected a 63% increase in reported domestic violence cases in the first quarter of 2020, as compared to the first quarter of 2019.[8] The increase in the number of distress calls strained the dedicated national domestic violence 107 hotline, which could only take a limited number of calls at a time. The police hotlines were also inundated with calls, both domestic violence- and non-domestic violence-related, during lockdowns, which was a barrier for domestic violence survivors to seek help. While some nongovernment organizations (NGOs) also had their own domestic violence helplines, these were only operational during regular work hours.[9] This was an issue especially considering that the number of distress calls came between evening and dawn.[10] Collectively, the police- and NGO-run helplines reported a 19% and 30% increase in calls, respectively, during the quarantine period.[11]

Despite the increase in calls, it is likely that more cases were unreported. Consultations with police and women revealed that many of the survivors chose to remain silent instead of seeking help because they were afraid of the possible consequences of reporting. These included losing economic stability in cases when the abuser is the household's breadwinner, possible reprisal, and backlash and social stigma after reporting (footnote 9). In addition, the movement restrictions brought by the pandemic limited the options available to survivors to seek help—as well as service providers to offer help—since the lockdown forced vulnerable family members to stay under the same roof as their abusers. ADB, through its

SUMMARY

What it is:
Artificial intelligence-powered chatbots were developed to help domestic violence survivors in Mongolia get assistance.

Start of implementation:
2020

Implementing department:
East Asia Department with support from the Department of Communication and Knowledge Management

Implementing partners:
- Ministry of Justice and Home Affairs
- National Legal Institute
- National Center Against Violence

Technology service provider:
Anduub Lab LLC

In line with ADB's operational priority:
- Accelerating progress in gender equality
  - Gender equality in human development enhanced
  - Women's resilience to external shocks strengthened

pioneering TA Addressing and Preventing Domestic Violence in Mongolia during the COVID-19 Crisis, worked with the National Legal Institute (NLI) and the National Center Against Violence (NCAV), a local NGO, to provide a continuum of care for

---

6   UN Women. n.d. *The Shadow Pandemic: Violence Against Women During COVID-19*.
7   International Development Law Organization. 2020. *Mongolia: Hand in Hand Against Domestic Violence*.
8   ADB. 2020. *Technical Assistance to Mongolia for Addressing and Preventing Domestic Violence in Mongolia During the COVID-19 Crisis*. Manila.
9   ADB. 2022. *Hands Talk: Duty of Care Continuum for Survivors of Domestic Violence in Mongolia*. 9 December.
10   ADB. 2020. *Preventing Domestic Violence Amid COVID-19 in Mongolia*. 25 November.
11   ADB. n.d. *Mongolia: Addressing and Preventing Domestic Violence in Mongolia During the COVID-19 Crisis*.

survivors in support of the country's landmark law on combating domestic violence (footnote 8). This marked the first time ADB helped increase the access of affected individuals to support services and prevent domestic violence during a crisis. The TA covered the development of digital tools to complement the existing technology infrastructure, such as hotlines, to provide more options for survivors to get help and lighten the load of the stretched police force. NCAV was already providing domestic violence services for many years but needed the digital platform to expand its services and reach more survivors and families during the pandemic.

Anduub Lab LLC, a local information technology (IT) company, was engaged to develop an AI-powered chatbot service. Given the sensitive nature of the cases, an interdisciplinary team of gender, IT, legal, and communication experts was formed. The team used the results of consultations with domestic violence survivors and frontline workers, such as counselors and police, to develop the content that was fed to the chatbot service to ensure that the information was accurate and relevant to those seeking help.[12] Stakeholder inputs also ensured that the chatbots reflected the language, terms, and contexts of the survivors to improve usability and care-seeking experience.

Subsequently, two chatbots, one for NCAV and one for NLI, were launched in April 2021 to offer users an expanded platform to safely get the information and services at the times they were needed most. The chatbots helped break down the barriers that prevent domestic violence survivors from seeking help by giving options for anonymity. Confidentiality and data privacy were major considerations to protect the identities of those who used the chatbots.

The NCAV chatbot provided information and services to ensure that domestic violence survivors were provided with a continuum of care, while the NLI chatbot provided legal information and gave referrals. Both chatbots were accessible on Facebook because consultations with the domestic violence survivors revealed high usage of the social media platform. The chatbots also had a mobile texting function to reach those who were based in rural areas and had limited internet connection.

The chatbot services also provided survivors with options for more in-depth, live counseling as needed. ADB covered the operational expenses of NCAV's live counselors who provided psychosocial and legal counseling for survivors who needed human conversation.

Survivors and families needing rescue and safe spaces were referred by the NCAV chatbot and live counselors to domestic violence shelters and one-stop service centers. Survivors continued to use the chatbot service even after rescue because they came to see these digital tools as their support system to move forward with their lives.[13]

Behavior change communication was an essential part of the continuum of care for survivors. A national multimedia campaign promoted the chatbots and their services and raised awareness about prevention and the many forms of domestic violence. The chatbots were equipped with information and services to change the behaviors of domestic violence survivors, from tolerating abuse as normal, to reporting, seeking support, and being economically independent. Providers were also trained to provide more sensitive first-response communication and handling of cases.

---

[12] ADB. 2023. *Technical Assistance Completion Report: Addressing and Preventing Domestic Violence in Mongolia during the COVID-19 Crisis*. Manila.

[13] R. Serafica, T. Begzsuren, and S. Gerelt-Od. 2023. Partnering with Civil Society to Combat Domestic Violence. *Development Asia*. 23 January.

By the end of the project in December 2022, NLI and NCAV received a total of 2,892 calls from over 3,000 users because of the chatbot service. An average of 14 new people used the chatbots between November and December 2022, when the media campaign was ramped up. The NCAV chatbot alone received calls from 1,187 new users (70.6% women, 24% men), while the NLI chatbot received 1,705 calls (68.2% women, 31.3% men) by December 2022. A total of 26 women and 16 children were rescued because of the chatbot service, while around 30% of the chatbot users were referred for in-depth psychosocial and legal services. The project endline showed that 90% of the users of the chatbot service reported that they were satisfied with receiving online domestic violence support and service.

The Ministry of Justice and Home Affairs has secured funding to continue supporting the chatbot services, while NCAV expressed its commitment to continue offering its services.

Driving Safe Transportation Using Digital Solutions

Public transport was heavily affected by COVID-19 not only because of the lockdowns that were imposed by governments across Asia and the Pacific grounded vehicles, but also because more people used private transport to protect themselves from infection. However, not all can afford to buy vehicles.

Mass public transport remains a key driver of the economies in the region. Thus, it was important to find a way to help the public transport sector adjust to the new normal. To this end, ADB launched the "Driving Safe Transportation Using Digital Solutions" challenge in August 2020 to crowdsource a digital solution to help public transport operators and users stay safe while traveling. ASafeRide, which was pitched by a team from Arup, was selected out of the proposals from 30 teams that participated in the challenge.[14]

ASafeRide was envisioned as a mobile application that uses crowdsourced data to provide passengers with real-time recommendations (e.g., what routes to take, which public transport vehicles are less crowded) to make their commute safer and faster. As such, it came with an e-wallet and payment option for cashless transactions. It was also envisioned to provide public transport operators with information to make vehicle deployment more efficient and responsive to demand. The app was intended to help resolve not only the issues brought about by the pandemic, but also the public transport problems that already existed even before COVID-19, such as unorganized fleets, unpredictable human and vehicle traffic, overcrowding in public transport vehicles, long queues, and arbitrary transportation fares.

The intent was to use AI, global positioning system (GPS), and big data to power the ASafeRide app. GPS could enable tracking the locations of passengers and public transport vehicles. The idea was to have data provided by public transport operators, along with publicly available information to be used by the app, to make route and vehicle recommendations to travelers. In addition, the data could be useful for contact tracing. The app also was meant to feature traffic calculation modules that could estimate travel times and lengths of queues.

SUMMARY

What it is:
ASafeRide was a mobile application that used big data, artificial intelligence, and global positioning system to provide real-time recommendations to public transport users to make their commute safer and faster.

Start of implementation:
2021

Technology service provider:
Arup

In line with ADB's operational priority:
- Addressing remaining poverty and reducing inequalities
  - Access to opportunities increased for the most vulnerable

ADB supported the development of the app's prototype. The support allowed the Arup team to fine-tune their concept, including clearly defining the problem they wanted to address, who their target market was, and what alternatives were available in the market that were similar to their envisioned product. The team also further refined the technical and digital components and the business model of their proposed solution. In addition, they consulted with stakeholders from Davao City (e.g., local government and public and private groups in the health and transport sectors) to get their feedback on the prototype.

A beta version was then developed based on the results of these consultations and the data. The beta prototype was designed to perform live vehicle tracking of jeepneys, which are a type of public transport vehicle in the Philippines, and provide users with estimates of their time of arrival when taking public transport. It was also intended to generate live data on the passenger count per public transport vehicle and automatically reflect changes in public transport options, e.g., routes and fare prices.

---

[14] ADB. 2020. Driving Safe Transportation Using Digital Solutions.

Restoring Public Confidence on Safe Travels

New protocols needed to be put in place to restore the public's trust in the safety of travel in 2021 following the global lockdowns that were put in place to prevent the spread of COVID-19. This was particularly important for countries in Asia and the Pacific that heavily rely on tourism for income generation. In 2019, nearly 10% of the region's gross domestic product (GDP) came from travel and tourism.

Innovation is necessary to help the public feel safe from infection even while traveling. To this end, ADB, in collaboration with various stakeholders, opened the "Restoring Public Confidence on Safe Travels" challenge to encourage digital solutions in order to help regain people's confidence in traveling.[15] Investo Medika Asia's proposed solution, MyHealth Diary, was selected out of 59 proposals. The app originally featured a database of various diseases and health-care facilities in Indonesia, but a new section on COVID-19 care was added in 2020. ADB supported the integration of the Health Passport feature, which was developed by the MyHealth Diary team in collaboration with several government institutions, including the Jakarta Provincial Health Department (Dinas Kesehatan DKI Jakarta) and the Jakarta City Transportation Council (Dewan Transportasi Kota Jakarta).

The Health Passport feature was intended to support efforts to track, trace, and monitor health conditions and inform users if they are safe to travel. The app was also intended to help users determine whether they were showing early symptoms so they could determine if they needed to get help. This feature was intended to help free up medical facilities, which were already overburdened with handling COVID-19 cases.

AI enabled the app to determine whether the user is safe to travel based on certain health indicators, e.g., temperature, heart rate, blood oxygen, and blood pressure. These indicators could be automatically collected by the wearables or manually entered on the app. Thresholds were set per indicator based on

medical standards to inform users whether they fell within the normal range, which was also shown on the app, or if their health condition may put them and/or other travelers at risk of getting sick. The app used the data to categorize users into one of three statuses: green for low risk/minimum risk, yellow for moderate risk, and red for high risk of health issues.

Daily data were also collected and visualized through a chart to enable users to track whether their health, through the indicators, has changed over several days.

The MyHealth Diary team secured a supplier that was able to meet its requirements, which included 100 wearables that were intended to be synchronized

SUMMARY

What it is:
MyHealth Diary uses artificial intelligence to enable users to monitor their health condition to determine whether they are safe to travel.

Start of implementation:
2021

Implementation partners:
Climate Change and Sustainable Development Department

Technology service provider:
Investo Medika Asia

In line with ADB's operational priority
- Fostering regional cooperation and integration
  - Greater and higher quality connectivity between economies
  - Global and regional trade and investment

---

[15] ADB. 2020. Restoring Public Confidence on Safe Travels.

with the app for users to easily monitor their daily health data. Meanwhile, the app development included ensuring that the data from the wearables can be synced to smartphones. The wearables were then distributed to app users who expressed interest in taking part in the testing. They provided feedback to the MyHealth Diary team on the wearables' performance. The application was also designed to prompt users to confirm and adjust the data in case of false detection, or when potential errors with the data were detected. Additionally, a feedback mechanism was also integrated to support the flagging of errors.

Users were also able to book and receive results of reverse transcription polymerase chain reaction (RT-PCR) or antigen tests using the application. They had the option to use the test results as supporting documentation to complement the health passport data. Other features of the COVID-19 care service included online booking for an RT-PCR test, ordering of medicine, and information on health facilities in Jakarta. Uploaded vaccination certificates on the app were accepted as supporting documents for entering public places. In effect, the Health Passport feature of MyHealth Diary served as a travel document for users within and outside the region.

In total, users received 100 wearables, but there were transfers of device ownership, leading to 150 individuals testing the Health Passport feature with wearables. Meanwhile, 470 individuals used the Health Passport feature without wearables.

The MyHealth Diary team partnered with institutions to promote their app, which was designed with a scanner feature to scan the quick response codes generated by the app to determine who would be allowed to enter based on their health status. The app was launched in Jakarta on 3 April 2021 through a drive-through COVID-19 vaccination activity at Jakarta International Equestrian Park. The event also served as an opportunity for the team to promote the app. The MyHealth Diary team used social media and tapped its existing network for promotion.

Predictive Model for Resettlement Outcomes

Field visits and desk reviews help ensure that land acquisition and resettlement (LAR) impacts and risks of ADB-supported projects are identified. Appropriate measures can then be developed and implemented to avoid or otherwise minimize, mitigate, and compensate for any and all adverse impacts.

Data-driven decision-making is necessary for resettlement planning and implementation. However, there were no models in place to forecast positive or negative outcomes of resettlement projects. In addition, there was no system in place to compare and analyze outcomes across projects. Inaccurate or delayed assessments due to data issues can affect compliance with safeguards requirements. Delays can also potentially negatively impact communities affected by projects.

This initiative saw the development of an AI-driven platform to carry out real-time resettlement monitoring and predict potential outcomes in Mongolia. Specifically, it tested whether machine learning could be used to predict resettlement outcomes using a combination of variables, including project land and acquisition requirements and baseline socioeconomic conditions of affected households.

This was the second phase of the testing of the platform that was developed by Mobiva, the technology service provider selected in ADB's "Real-time Tracking of Resettlement Implementation" challenge.[16] The team was originally tasked with developing a proof of concept (POC) of a dashboard that allowed users to collect and share real-time resettlement data via a digital platform. The POC also illustrated how it could be used to visualize resettlement information.

The initiative, which was done under the Mongolia Ulaanbaatar Urban Services and Ger Areas Development Investment Program Tranche 1 and Tranche 2, was the first time for the bank to test a machine learning model for post-resettlement

SUMMARY

What it is:
This initiative tested whether machine learning could be used for forecasting resettlement outcomes.

Start of implementation:
2022

Implementation partners:
- East Asia Department
- Government of Mongolia
- Project Management Office of the Mongolia Ulaanbaatar Urban Services and Ger Areas Development Investment Program

Technology service provider:
Mobiva

In line with ADB's operational priority:
- Strengthening governance and institutional capacity
  - Strengthened country systems and standards

evaluation. This solution was in line with ADB's East Asia Department's digital transformation agenda and the Government of Mongolia's E-Mongolia policy (Vision 2050).

The key activities were as follows:

Data matching and cleaning. The pre- and post-LAR datasets were first matched to ensure that the same households were compared. The Tranche 1 data resulted in 144 entries that were successfully matched. Entries with duplicate and null values were also removed as part of data cleaning.

---

[16] ADB. 2020. Real-time Tracking of Resettlement Implementation.

Data pre-processing and normalization. The categorical data were converted into numerical data to make it suitable for machine learning algorithms. Data normalization was then done to mitigate any bias due to scale disparities.

A histogram based on the available data was also created to show the changes in monthly household income from pre-LAR to post-LAR.

Model design and training. Different AI models were developed and tested based on the Tranche 1 household data: decision tree, artificial neural networks, and k-nearest neighbor. Testing showed that the artificial neural network model was better at making predictions than k-nearest neighbor and decision tree, particularly for the "No change" and "Increase" categories. Three AI models were developed: household income, household poverty, and household satisfaction. The performance of these models in forecasting resettlement outcomes was tested using Tranche 2 data.

The Tranche 2 datasets were also pre-processed and cleaned to ensure data quality and consistency. Some discrepancies were observed in the pre- and post-LAR datasets during this stage. Specifically, "lighting system" and "cooking fuel" were not present in the Tranche 2 pre-LAR data set, so these were replaced with "power source" and "heating energy source" that were included in the data set. There were also differences in the content for "source of drinking water" and "latrine condition."

The clean Tranche 2 data were used to test whether the model was able to accurately forecast resettlement outcomes. The F-1 score, which measured precision and recall, was used to assess the models' performance. The results were as follows:

- **Household income**. The model showed significant improvements in precision, recall, and accuracy, indicating the robustness in the model's ability to predict changes in household income.

- **Household poverty**. While recall improved in Tranche 2, both accuracy and precision dropped in Tranche 2, which may be due to the higher rate of false positives.
- **Household satisfaction**. Precision, recall, and accuracy all dropped in Tranche 2, raising questions about the model's ability to capture nuanced factors that affect household satisfaction. Another possible reason for the significant decrease across all indicators was the replacement of "lighting system" and "cooking fuel" with "power source" and "heating energy source," respectively, which may have affected the model's forecasting ability.

The models for predicting changes in household income and household poverty have the potential to be used in future resettlement projects, subject to additional training and testing using larger datasets to improve accuracy.

The model used to predict household satisfaction, on the other hand, may need to be revisited considering that the Tranche 2 pre- and post-LAR datasets changed. Additional testing may be done to see whether the results would still be the same if the pre- and post-LAR questions remained consistent. The models can be trained and tested using the same datasets consistently from the start to the end of the resettlement process.

Robo-Advisor

ADB, with the support of Oracle Labs, developed the POC of Robo-Advisor to broaden its capacity for information dissemination by making it easier for both internal and external audiences to learn about the bank's initiatives. This was intended as one of the initiatives to promote digital development and innovative technologies, one of the operational approaches identified in ADB's Strategy 2030.[17] It was also in line with the Digital Agenda, which supported exploring new and emerging technologies to help address business challenges and improve the bank's agility and responsiveness.[18]

Originally, the POC was only intended to come in the form of a chatbot that answered text-based queries. However, it was later decided to explore whether the digital assistant could be integrated with Pepper, a social humanoid and programmable robot, to complement the text-based chatbot app with a voice user interface. The intent was to explore the possibility of using Pepper during events so guests can get more information about ADB's projects and initiatives in an interactive manner.

AI was used to train the Robo-Advisor POC. Publicly available documents in the projects section on ADB's website were used in the training, covering three thematic areas: climate change, gender, and finance. Meanwhile, an intermediary service was used to integrate Robo-Advisor with Pepper. This was tested during the Third Asia Finance Forum in November 2019.

The chatbot version would have been beneficial for ADB's international staff and clientele, who come from several countries with different time zones. Having a tool that would readily answer queries would have also helped build trust among ADB's various stakeholders as they would have been able to familiarize themselves with the bank's operations without needing to sort through thousands of documents or consult with ADB staff and consultants to find the answer to their specific queries. Another benefit that the tool offered to ADB was its capacity

SUMMARY

What it is:
A digital assistant accessible through a chatbot and a physical robot that answered queries related to ADB's operations concerning climate change, gender, and finance by sifting through publicly available documents.

Start of implementation:
2018

Implementation partner:
Climate Change and Sustainable Development Department

Technology service provider:
Oracle Labs

In line with ADB's operational approaches:
- Promoting digital development and innovative technologies
- Developing capacity
- Adding value and promoting quality infrastructure

to provide consistent responses, unlike humans who may provide different answers depending on their training and experience.

However, various issues affected the POC development. PDF files uploaded on ADB project pages had to be converted into other file formats (e.g., .txt, .xls, .json) and had to be manually reviewed and validated. This limited the data used for training the model. Only 8,000 to 10,000 documents were considered out of the 60,000 documents available on the website at the time. Questions for Pepper

---

[17] ADB. 2018. *Strategy 2030: Achieving a Prosperous, Inclusive, Resilient, and Sustainable Asia and the Pacific*. Manila.
[18] ADB. 2018. *Digital Agenda 2030: Special Capital Expenditure Requirements for 2019–2023*. Manila.

ADB ASIAN DEVELOPMENT BANK

Intelligent Search and Digital Assistant

| Examples ▼ | When was MFF approved? | | | Search |

Search Results

| Document Span / Text | Doc. Score | Span Score |

doc1

A Multi tranche Financing Facility (MFF) was approved on **15 September 2006** for $430 1.million to improve the performance of the railway sector and to increase the capacity of the existing rail network to handle traffic demand necessary to sustain Bangladesh's economic growth.

6 — 1

ADB's Approach to the Climate Change Challenge

ADB's approach to addressing climate change is articulated in ADB's Climate Change Operational Framework 2017–2030 which provides a vision of enhanced actions for low greenhouse gas emissions and climate-resilient development in the Asia and the Pacific. ADB's work is guided by the needs of its members and their commitment to global agendas (e.g., Paris Agreement, Sustainable Development Goals, and Sendai Framework or Disaster Risk Reduction), with the aim to provide the needed assistance in terms of financing, capacity development, technology development and knowledge enhancement. In **July 2018**, ADB approved its new institutional strategy, the Strategy 2030, with tackling climate change, building climate and disaster resilience and enhancing environmental sustainability as one of its seven operational priorities (OPs). This OP (or OP3) focuses on: (i) scaling up support to address climate change, disaster risks, and environmental degradation; (ii) achieving ADB's climate operations target; (iii) accelerating low greenhouse gas emissions development; (iv) ensuring a comprehensive approach to build climate and disaster resilience; (v) ensuring environmental sustainability; and (vi) increasing focus on the water-food-energy security nexus. ADB is the first multilateral development bank (MDB) to set clear climate investment targets, aiming for 75% of the number of ADB's committed operations (on a 3-year rolling average, including both sovereign and nonsovereign operations) supporting climate change mitigation and adaptation by 2030; and climate finance from ADB's own resources to reach $80 billion cumulatively from 2019 to 2030.

2 — 6

How can we help? ✕

ADB Digital Assistant Bot
We're here to talk, so ask us anything!

Type a message...   Send

Sample response from Robo-Advisor.

during demonstrations also had to be simple as the voice user interface was unable to answer long and complex queries.

The advent of the COVID-19 pandemic halted activities related to this initiative. While this was parked, rapidly evolving technology outside of ADB led to the development of better options on the market that are available out of the box, such as ChatGPT, removing the need to develop a custom model for ADB.

The "One ADB" approach explicitly states that ADB aims to bring together knowledge and expertise to effectively implement Strategy 2030.[19] Thus, it was essential for the bank to have a system in place that allowed users to find the right information quickly and easily to support decision-making. This was particularly important considering that DMCs are now experiencing increasingly complex issues, including those brought by climate change. Identifying common themes and lessons may be useful to policymakers who face similar issues.

The Independent Evaluation Department (IED) is responsible for reviewing and extracting key lessons from ADB documents to feed into sector, corporate, and thematic evaluations and support the design of new operations. In the past, it had no platform in place that automatically generated and curated insights to provide users with information to guide them in decision-making. Instead, IED staff manually reviewed these documents to extract lessons. This was a challenging task considering that ADB's institutional knowledge is spread across over 20,000 documents, including project completion reports and project validation reports, with more added yearly. It also has research products that provide policy advice and different perspectives across Asia and the Pacific. Typically, staff needed half a day to review and extract lessons from one document.

A keyword-based enterprise search function was not sufficient for a platform intended to extract lessons from key ADB documents, especially since this traditional search tool heavily relied on the user knowing what keywords to use to get correct results. What was needed was a new intuitive system that could cull valuable insights quickly and easily from documents. Thus, ITD and IED collaborated to develop EVA, a cognitive search tool.

A major task that had to be undertaken was to "teach" EVA to identify what are and are not lessons. ADB adopted the definition of "lesson" used by the International Labour Organization:

SUMMARY

What it is:
A digital tool that leveraged artificial intelligence to automate the extraction of knowledge and lessons from ADB documents, including project completion reports and project validation reports.

Start of implementation:
2019

Implementation partner:
ADB's Independent Evaluation Department

In line with ADB's operational approaches:
- Promoting digital development and innovative technologies
- Delivering integrated solutions
- Adding value and promoting quality infrastructure

"A lesson learned is an observation from project or programme experience which can be translated into relevant, beneficial knowledge by establishing clear causal factors and effects. It focuses on a specific design, activity, process, or decision and may provide either positive or negative insights on operational effectiveness and efficiency, impact on the achievement of outcomes, or influence on sustainability. The lesson should indicate, where possible, how it contributes to 1) reducing or eliminating deficiencies; or 2) building successful and sustainable practice and performance."[20]

Annotations also needed to be made to improve the accuracy of EVA. Project completion reports were initially loaded to the model, followed by

---

[19] ADB. 2022. *One ADB: An Evaluation of ADB's Approach to Delivering Strategy 2030*. 8 February.
[20] International Labour Organization. 2014. *Evaluation Lessons Learned and Emerging Good Practices*.

project validation reports, TA completion reports, and TA completion report validation. IED staff were tasked with annotating and validating reports to provide initial inputs to the model as part of the teaching process.

The minimum viable product (MVP) was officially launched and identified for scaling in 2020. EVA identifies, generates, and ranks lessons according to importance and relevance based on what users searched for. The tool extracts lessons from thousands of ADB documents quickly, allowing users, particularly IED staff, to save time.

Users, meanwhile, can filter results by sector, country, theme, and year. They can also correct wrongly tagged lessons to help train the model to recognize what are and are not lessons.

The full version of EVA, once rolled out, is expected to offer numerous benefits, including making evaluation knowledge more accessible and used, increasing efficiency in reviewing and evaluating ADB documentation, and making information available on demand to support decision-making.

CHAPTER 2

Artificial Intelligence in Economics

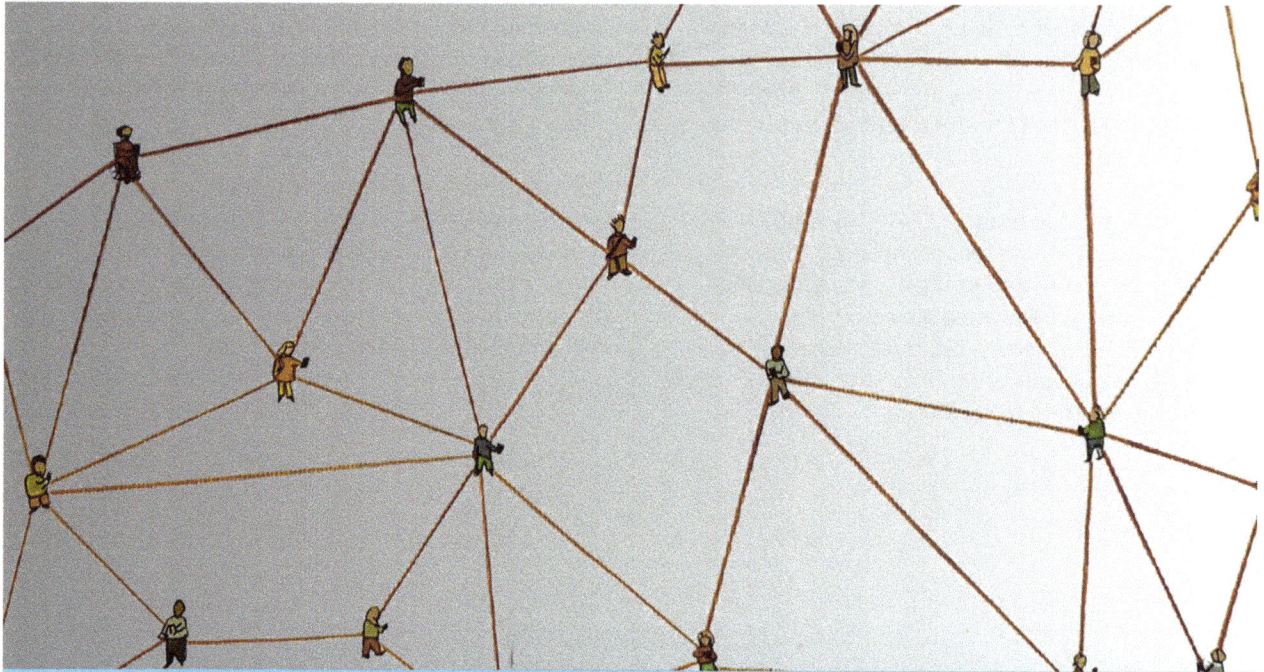

Technology can be used to connect people.

Using AI is not new in the field of economics. It has long been used by experts for data crunching and predicting market trends. Technological advancements, however, have allowed people to extract more meaning from the data, enabling them to get better insights on the relationships of various economic and financial indicators.

This section features the AI work ADB has done in the areas of economic nowcasting and forecasting. Some of these initiatives were developed because of the COVID-19 pandemic, which triggered significant delays in the collection and release of economic data. To remedy that, various proxies, including night lights, social media, and newspaper articles, were analyzed to see if they would generate information on DMCs' economic activities.

With the advent of generative AI, more tools and innovations will likely be developed to get more meaningful insights to aid decision-makers in developing economic policies.

Economic Monitoring Using Alternative Data Sources

A country's GDP is an important indicator of the size and health of its economy.[21] It is considered by policymakers, investors, and other stakeholders in policy formulation and decision-making. However, there is usually a delay in the release of quarterly GDP, particularly for many developing countries that have sparse data.

A growing number of emerging new technologies and alternative data sources can be used to generate insights on a country's economic status within a given period more quickly compared to GDP. However, the sheer volume and complexity of the information makes it difficult for human users to sift through and interpret. In addition, big data may also not be compatible with traditional econometric models.[22] New models need to be developed to make sense of big data for nowcasting.

The initiative explored the possibility of using news articles to augment macroeconomic data for nowcasting. To do this, three models were developed.

Quantitative data was used to build the economic indicator model (EIM) using the monthly records for 44 indicators provided by ADB, along with seasonally adjusted data. Meanwhile, textual data from news articles were used to develop the thematic model to compensate for the economic data availability issues in EIM. A web scraping tool for a news aggregator was used to collect a total of 2.2 million articles released by various news outlets from October 1989 to December 2019. Various publishers were included to minimize bias and normalize the writing styles of different authors. Data cleaning, followed by lemmatization and token generation, was then performed. Keyword co-occurrence networks were also built to show the relationship strengths between keyword pairs. The most frequent unique tokens in an article were used as the keywords of the article to reduce noise and lessen the strain on the machine when performing network analysis and modeling. Various algorithms were then developed and trained to model the GDP growth rate using the keywords.

SUMMARY

What it is:
Information from news articles was mined using artificial intelligence and network science to make timelier and more accurate nowcasts of economic activities in the Philippines.

Start of implementation:
2020

Implementation partners:
- Economic Research and Development Impact Department
- Asian Institute of Management

In line with ADB's operational approach:
- Promoting digital development and innovative technologies

EIM and the thematic model were independently developed, enabling the team to generate these models in parallel. Network analysis was also conducted to capture dynamic relationships and uncover themes and topics.

A third model, called the theme augmented model (TAM), combined the data sources of the two other models to offer a more holistic picture of the GDP, compensate for the respective weaknesses of the EIM and the thematic model, and produce more accurate predictions.

Three approaches were used to develop the TAM. The first model, dubbed unified TAM or U TAM, was an XGBoost model that used both economic and textual features. The U TAM did not offer much of an improvement to the thematic model regarding the timely generation of accurate nowcasts. The second

---

[21] International Monetary Fund. n.d. *Gross Domestic Product: An Economy's All.*

[22] D. Hirschbühl, L. Onorante, and L. Saiz. 2021. Using Machine Learning and Big Data to Analyse the Business Cycle. *Economic Bulletin*. 5. Frankfurt: European Central Bank.

approach, dubbed E1 TAM, was a Random Forest regressor that used the predictions of the EIM and thematic model. The third approach, an ensemble model called E2 TAM, incorporated predictions from U TAM with the EIM and thematic model results.

All the models were trained using a common target and evaluated using the same metrics and methodology. The training set covered data from 2000 to 2019 and was used for hyperparameter tuning to minimize root mean squared error. Meanwhile, the results were compared to common baselines, which included the Bloomberg consensus forecasts, an autoregressive random walk, and a naïve assumption, which mimicked how GDP growth rates were reported in press releases relative to previous year values. Multiple consecutive test samples were identified, with each of the models trained using data before each quarterly sample in a process the team called "extending window training and evaluation" to reduce the mean absolute error. For instance, GDP data from the first quarter of 2000 to the fourth quarter of 2015 was used to check whether the model was able to correctly predict the GDP of the first quarter of 2016.

The exploratory initiative showed that news articles are useful in generating accurate and timely nowcasts: nowcasting with news alone can reduce the margin of error to 0.36 percentage points. Being able to use a combination of methods and tools—including word co-occurrence networks, correlation networks, and community detection— with machine learning for GDP nowcasting can aid key decision-makers in government, nongovernment, and finance organizations in making decisions related to areas such as poverty alleviation, labor, trade, and investments.

Using the Automatic Identification System to Generate Insights on Maritime Activities

The Automatic Identification System (AIS) is an automated tracking system used at sea that provides information on ships within a given time.[23] While it was originally designed to help ships navigate and avoid collisions, its data are now increasingly being used for research, port performance analysis, and estimates of trade flows and maritime carbon trade emissions.

ADB explored whether AIS data, which are available through the United Nations Global Platform, could be used as an alternative source of economic statistics, considering that it is available in near real time (data are updated every 4 hours), while official statistics take months or years before they are released.[24]

The key indicators that were identified were not explicitly identified in the raw AIS data. Thus, ADB proposed a framework and methods to extract the information from the data. ADB's proposed framework covered events of interest (EOIs), which are specific maritime incidents or activities pertinent to a target indicator, and areas of interest (AOIs), which are geographic locations where these events occur identified through manual, distance-based, or cluster-based approaches. The framework was operationalized by creating indicators for ports and passageways that represent major hubs of maritime activity and with cases of maritime disruptions:

1. **Port activity.** Three indicators were identified under port activity: the count of unique vessels in a port, the number of arrivals in a port, and the median time spent by vessels in a port.

   The EOIs were the port calls (entry, length of stay, and exit of vessels to and from the port) and the AOIs cover the berths, terminals, and anchorages of each port. Squares that were formed by tracing 22 km from the center of each port for every side served as the boundaries to mark the distance-based AOIs.[25] These were supplemented with

SUMMARY

What it is:
ADB tested whether the Automatic Identification System, which is used to support ship navigation, could serve as an alternative data source to generate insights on maritime activities.

Start of implementation:
2022

Implementing department:
Economic Research and Development Impact Department

In line with ADB's operational approaches:
- Promoting digital development and innovative technologies
- Applying differentiated approaches

cluster-based AOIs formed using Hexagonal Hierarchical Geospatial Indexing System (H3), and Density-Based Spatial Clustering of Applications with Noise (DBSCAN). The H3 indices were used to represent the location points while the DBSCAN algorithm identified the group of H3 indices forming the port boundary.

2. **Traffic along maritime highways.** Three indicators were identified: the number of unique vessels, the count of transits, and the time spent by the vessels in these passageways.

   The EOIs were the vessels' entry and exit from these passageways. The AOIs, meanwhile, were the mouths of the passageways identified using

---

23   UN Big Data. 2024. AIS Data: Task Team of the UN Committee of Experts on Big Data and Data Science for Official Statistics.
24   This case study is largely based on the 2023 publication that presents the framework and study on AIS data; ADB. 2023. *Methodological Framework for Unlocking Maritime Insights Using Automatic Identification System Data*. Manila.
25   The United Nations Convention on Laws of the Sea states that a country's territorial zone is 22 km from its low-water coastline.

DBSCAN. Manual AOIs generated by selecting the narrowest areas along the passageways were used as alternative AOIs when it was not feasible to use the mouths.

To study the major hubs of maritime activity, the largest ports identified by the World Shipping Council in 2019 and those with the highest connectivity to different parts of the world were selected. These included the ports of Los Angeles and Long Beach in the United States, the Port of Rotterdam in the Netherlands, and the Port of Shanghai in the People's Republic of China.

Traffic in the passageways was supplemented to port activities to create a comprehensive view of the maritime industry. Passageways included were the Malacca and Singapore Straits, the Suez Canal, the Strait of Gibraltar, and the Panama Canal. The straits of Hormuz and Bab-el Mandeb were also included considering the role they play in the global oil trade.

Ports and passageways that faced significant disruptions in 2022 (the Russian invasion of Ukraine, the Sri Lankan economic crisis, and the Tonga volcanic eruption) were also studied to see whether AIS data significantly changed because of these events. The ports and passageways that were included in the research because of this consideration were the Port of Odesa, the Port of Colombo, the Port of Nuku'alofa, the Dardanelles Strait, and the Bosporus Strait.

The indicators were able to capture the effects of the lockdowns due to the COVID-19 pandemic, which saw reduced maritime activity. It also reflected the uptick in activity by February 2022. Similarly, it also revealed the impact of the Russian invasion of Ukraine on trade activities. Meanwhile, the blockage of the Suez Canal in March 2021 was also captured by the indicators. On the other hand, the study also revealed some limitations in using AIS data, including signal gaps in cases when transponders are turned off.

Some of the indicators were validated by comparing them with other data sources. While there were some discrepancies, the estimates were nonetheless coherent with official data, with absolute percentage errors ranging from 3%–4%, which lends credence to the value of using AI data to generate timely and accurate statistics.

The results of the study showed that AIS data could be used as an alternative data source to generate timely information on maritime activities. It is important to note that the indicators were developed not to replace official statistics but rather to complement them by providing near-real-time insights to help policymakers make data-driven decisions.

Nowcasting the Economic Impact of COVID-19

The COVID-19 pandemic has become one of the biggest global health crises, but its impact goes beyond health. Experts project that the poorest countries will feel the impact of the pandemic for years,[26] while ADB warned in 2020 that the global economy could suffer up to $8.8 trillion in losses because of the pandemic. Asia and the Pacific account for about 30% of the overall decline in global output, which is equivalent to economic losses worth $2.5 trillion.[27]

Policymakers and businesses need timely data to base decisions that would support economic recovery. Official statistics are released on a monthly, quarterly, or yearly basis. However, some developing countries experience a significant lag in releasing this data; it can take any time between a few weeks to several months for these to be released.

This initiative saw the development of an MVP that could demonstrate how multiple and varied datasets could be used for nowcasting to determine COVID-19's impact on GDP growth rate in select countries in Asia and the Pacific.[28] The MVP was developed by Ddata, which was selected in ADB's "Nowcasting the Economic Impact of COVID-19," besting 65 other teams.[29] Ddata used a combination of big data and machine learning to develop its nowcasting model that links and aggregates thousands of datasets to predict the movement of a target indicator, in this case, GDP growth rates. Raw data came from multiple sources, had different types, and had varying release dates to enable the model to reflect holistic economic measurements, eliminate noise, and increase the model's accuracy in making predictions.

The solution offered by Ddata had two main features:

- **Data ingestion, transformation, and modeling.** Ddata reformatted various datasets into a common format and linked them together.

SUMMARY

What it is:
Artificial intelligence and big data were used to extract relevant information from large datasets to predict gross domestic product growth rates.

Start of implementation:
2021

Implementation partner:
Economic Research and Development Impact Department

Technology service provider:
Ddata

In line with ADB's operational priority:
- Strengthening governance and institutional capacity
  - Strengthened public management and financial stability
  - Enhanced governance and institutional capacity for service delivery
  - Strengthened country systems and standards

All data transformations were recorded for transparency. Quality check was also performed for maximum accuracy. Ddata combined linear and nonlinear methods (i.e., neural networks) to develop the model.
- **Backfilling.** Ddata trained and tested the model using data from the 2008–2009 financial crisis because this period, like the COVID-19

---

26   A. Gulland. 2021. Long-term Impact of Pandemic Will Be Felt for Decades, Warns Red Cross. *The Telegraph*. 22 November.
27   ADB. 2020. COVID-19 Economic Impact Could Reach $8.8 Trillion Globally—New ADB Report. News release. 15 May.
28   Nowcasting is done to obtain early estimates of key economic indicators, which are usually released at low frequencies and with significant delays, by using data related to the identified variables but are collected more frequently and released in a timely manner; A. Stundziene et al. 2023. Nowcasting the Economic Activity Using Electricity Market Data: The Case of Lithuania. *Economies*. 11 (5). 134.
29   ADB. 2020. Nowcasting the Economic Impact of COVID-19.

pandemic, was marked by economic volatility. This was done to check if it could accurately predict the economic shocks and recovery using data from that period. In cases where data was missing, "backfilling" was done using machine learning to fill in the missing data points using the available variables and their relationships.

ADB and Ddata agreed to narrow the focus of this initiative to the quarterly GDP growth rates of five countries in the Asia and Pacific region (Armenia, Fiji, Myanmar, Pakistan, and the Philippines). These countries had different geographical locations and varied in terms of the types of data they release and the frequency with which these official statistics are released, making them good representative cases for the Asia and Pacific region.

Ddata presented its final output to ADB on 29 July 2021. The model accurately nowcasted GDP growth over time for 6–12 months across all countries using multiple data sources.[30] The accuracy was particularly good for high-data countries such as the Philippines. Meanwhile, the model was able to identify data-point variations for low-data countries such as Pakistan and Myanmar. The nowcast results were released weekly, thus ensuring timeliness.

The model was in line with or even better than similar models in terms of analysis errors. Ddata also developed cloud-based automated pipelines that allowed the creation of release-based updates.

The nowcasting model has progressed to the beta stage of testing by the end of the MVP development. Ddata intended to gather more data to further improve its model, which is scalable, by incentivizing people to add data to the peer-to-peer distributed ledger technology-based data-sharing platform they created to crowdsource alternative datasets.

---

[30] Some of the data types and sources used in the study are as follows: (i) governmental—includes accounts, production, sectors finance, interest rates, investment trade, e.g., CEIC data, Penn World Tables; (ii) academic—covers poverty, competitiveness, governance, and complexity, e.g., World Bank Development, World Economic Forum Global Competitiveness Index, International Monetary Fund World Economic Outlook; (iii) financial stocks—covers stock prices, e.g., Armenia Securities Exchange, South Pacific Stock Exchange, Philippines Stock Exchange; (iv) financial index—covers global indexes, commodity prices, trade index, exchange rates, global bond volatility, and equity, e.g., NASDAQ Global, Yale S&P, MarketWatch; (v) sentiment—covers news sentiment and investor confidence, e.g., FnSentS Web News Sentiment, Yale Confidence, AAII Investor Sentiment; and (vi) sensor—covers information on weather, COVID-19, and traffic, e.g., Open Weather data, CEIC.

Earth Observation Data

Various developments in the past years, including additional investments in satellite capabilities, advancements in algorithms and data processing tools, and increasing data accessibility, have led to the increasing popularity of using earth observation data to generate insights.[31] COVID-19 further accelerated its use as more governments and organizations are exploring alternative data sources given that the pandemic affected the collection and sharing of official statistics.

Earthlab AI Systems, which was the team selected in the "Earth Observation Data Challenge," explored whether it was possible to use different sources of data collected by satellite missions for correlating and predicting economic activities.[32] The original intent was to have an integrated dashboard for monitoring multiple economic indicators across different countries, but it was later decided that the initiative would instead focus on investigating whether there were correlations between proxy indicators using earth observation data and economic activities in three countries: Georgia, the Philippines, and the Republic of Korea.

Besides remote sensing and geographic information system mapping, the team used AI, specifically machine learning, and big data to make sense of the earth observation images taken by satellites. The amount of earth observation data collected by satellites is huge. Making sense of the information can be challenging given the extent and capacity of humans to interpret data.[33] Machine learning can be used to detect patterns and similarities in large amounts of data.

Four approaches were used:

1. **Ships in ports.** The presence of ships in ports was used as a proxy for trade volume from international shipping. Synthetic aperture radar images from the European Space Agency Copernicus Mission Sentinel 1 were analyzed since radar satellite data are not affected by weather conditions such as clouds. Areas that showed highly significant

SUMMARY

What it is:
Data extracted from satellite images were analyzed to see if these could be used to correlate and predict economic activities.

Start of implementation:
2021

Implementation partner:
Economic Research and Development Impact Department

Technology service provider:
Earthlab AI Systems

In line with ADB's operational priorities:
- Fostering regional cooperation and integration
  ‣ Global and regional trade and investment opportunities expanded
  ‣ Regional public goods increased and diversified
- Strengthening governance and institutional capacity
  ‣ Strengthened public management and financial stability

positive correlations between the number of ships in their ports and economic indicators included the Port of Davao and the Port of Gwangyang. The Port of Busan, meanwhile, showed a negative correlation between the average monthly number of ships in the port and exports as an economic indicator. Other factors could have affected the presence of ships in ports. For instance, the movement restrictions in the first half of 2020 saw an increase in the number of ships in the Port of Manila because of quarantine restrictions.

---

[31] Organisation for Economic Co-operation and Development. 2017. *Earth Observation for Decision-making*. Paris.
[32] ADB. 2020. *Earth Observation Data Challenge.*
[33] D. Tuia et al. 2023. *Artificial Intelligence to Advance Earth Observation: A Perspective*.

2. **Nitrogen dioxide concentrations.** Nitrogen dioxide is emitted when fuel is burned, and as such is associated with human activity. The European Space Agency's Sentinel-5p L2 measures nitrogen dioxide concentrations globally, alongside other variables. There was also a clear seasonal effect on the nitrogen dioxide concentrations across all three countries. While all calculated correlation coefficients were positive and some were statistically significant, there were inconclusive results on whether there was a significant correlation between nitrogen dioxide measurements and economic indicators. This may have been due to various reasons, including the complexity of the relationship between nitrogen dioxide and different economic indicators and meteorological conditions (such as precipitation and wind). It is also possible that the time series of 3 years was too short.

3. **Containers in ports.** The number of containers in ports was also studied as a possible proxy indicator for trade volume. Images from PlanetScope, which takes images of the land surface of the earth every day using approximately 130 satellites, were analyzed. The small sample size may have contributed to inconclusive results. There was a strong but not significant negative correlation between the number of containers and the import and export activity. This may have been affected by the port's space utilization. For instance, containers may have piled up when trade slowed down.

4. **Nighttime lights.** Nighttime lights can provide insights on areas that had limited available data on socioeconomic indicators.[34] The working assumption in this initiative was that there was a correlation between nighttime light and industrialization in a given area. However, there were inconclusive results. Further research may be needed to determine the relationship between economic activities and nighttime light. It is possible that the available instruments at the time the satellite images were collected were not sensitive enough to capture the subtle differences in nighttime light. Additional calibration, cleaning, and processing of the data may be needed as the available images did not seem to show significant differences between the average pixel radiance per month. There may also have been other variables that influenced the relationship between economic activity and nighttime light.

This initiative showed a lot of potential but time, resources, and technical support from experts are needed to explore these approaches further. Nonetheless, these results may be used as a starting point to pursue additional economic analyses. Similar studies should consider unique geographical and socioeconomic contexts. A clear understanding of the tools and technologies is also important to ensure correct interpretation and analysis. Care must be taken in analyzing and forming conclusions considering that some of these approaches used data that covered limited time periods. Data that cover longer periods can provide a better understanding of the relationships of different variables.[35]

34   A. Martinez et al. 2016. How Nighttime Lights Help Us Study Development Indicators. *Asian Development Blog.* 22 December.
35   ADB. 2023. Appraising New Damage Assessment Techniques in Disaster-Prone Fiji. *ADB Briefs.* No. 240. February.

CHAPTER 3

Artificial Intelligence in Administration

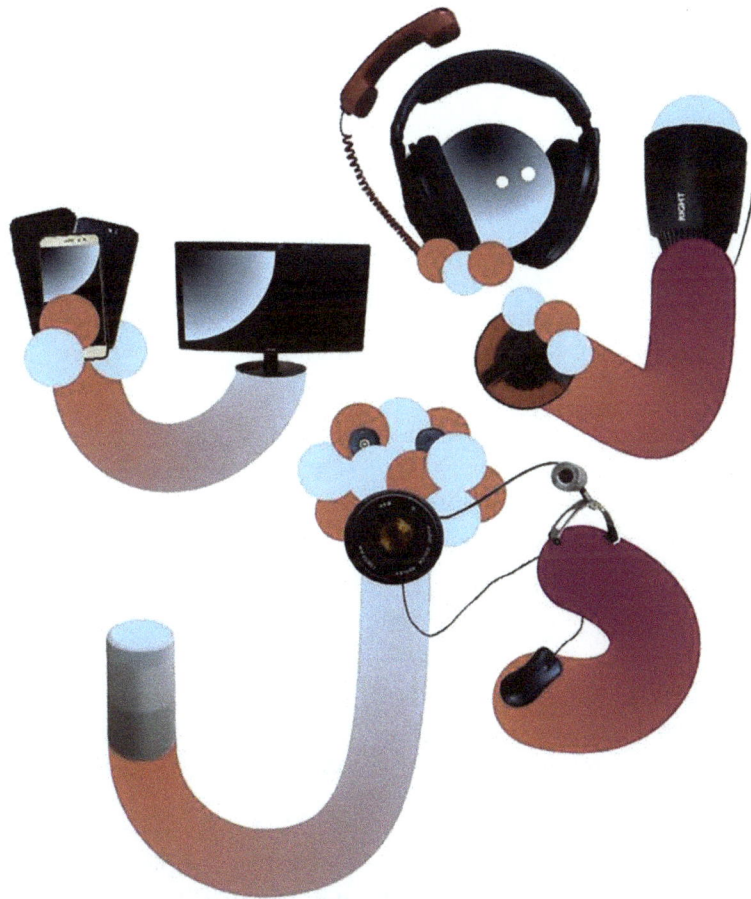

AI can be used to augment and support humans in the workforce.

Image by Anton Grabolle / Better Images of AI / Human-AI collaboration / CC-BY 4.0

Much has been said about how automation is revolutionizing workplace management. However, there are other ways in which AI can be used to transform office administration. This section looks at ADB's exploratory efforts to use AI to improve efficiency at work. It is worth noting that some of the earliest AI tools developed by the bank were designed with this in mind. MyADB Recruitment Intelligence was developed to see how AI could streamline the recruitment process, while the Intelligent Integrity Due Diligence Platform was intended to test how the technology could be used to perform due diligence checks. Technical Assistance Messaging Intelligence, meanwhile, can provide updates on disbursement-related concerns. The Code of Conduct Chatbot, on the other hand, was designed to guide personnel so they can uphold ADB's standards on ethics and integrity.

The increasing number of digital assistants making their way to ADB shows that the digital workforce is here to stay to provide support to human personnel.

Code of Conduct Chatbot

The Office of Professional Ethics and Conduct (OPEC) is responsible for ensuring that all ADB personnel perform their functions according to the standards of ethics and integrity upheld by the bank. To help achieve this, OPEC and ITD collaborated to develop the Code of Conduct Chatbot, which is a communication tool trained to answer queries based on ADB's Code of Conduct, as well as other key documents. The chatbot can be accessed on computers or mobile phones via Microsoft Teams for users' convenience. It is also accessible 24/7 and supports various languages, in consideration of the fact that personnel—regardless of whether they are from headquarters, resident missions, field offices, or the ADB Institute—come from different countries and time zones. The confidentiality offered by the tool also helps OPEC engage with and build the trust of personnel, including those who are apprehensive or fearful of asking sensitive questions.

The chatbot also offers other benefits. It works independently, freeing up OPEC to focus on key tasks such as the facilitation of workplace concern resolutions. Furthermore, it serves as a strategic tool: the insights OPEC gets from the digital tool enable the office to tailor its training and outreach efforts to address the concerns of ADB personnel. In effect, the digital tool helps OPEC optimize its operations. At the same time, it also makes it easier for personnel, particularly those who are new to ADB, to find the information they are looking for quickly and easily. Besides answering questions, the Code of Conduct Chatbot also cites the relevant provisions and provides the hyperlinks of the source documents should users need additional information.

Given the nature of the topics handled by OPEC, care was taken to ensure that the answers the chatbot generates only come from the data it was fed during training. This was done so that the chatbot maintains consistency and accuracy in its responses. If necessary, it asks questions in case of unclear queries. In addition, the chatbot reminds users that it is not allowed to give authoritative advice and refers them to OPEC for queries that the chatbot is unable to answer.

SUMMARY

What it is:
An artificial intelligence-powered chatbot designed to answer queries related to ADB's rules and guidelines on ethics and integrity.

Start of implementation:
2023

Implementing department:
Office of Professional Ethics and Conduct

In line with ADB's operational approach:
- Promoting digital development and innovative technologies

ITD and OPEC monitored the performance of the app during the pilot stage to ensure that answers were accurate and there were no errors.

The Code of Conduct Chatbot was officially launched in October 2023 and is the first chatbot in ADB to be powered by GPT 4. A communication campaign was rolled out to inform staff and stakeholders on various channels and activities that the digital tool was now available for use.

Technical Assistance Messaging Intelligence

The Controller's Department Loan Administration Division–Technical Assistance Section (CTLA-TA) authorizes TA-related disbursements in ADB. It reviews the submitted claims and supporting documents from internal clients (Technical Assistance Supervising Units or TASU) and external customers (including TA consultants and service providers) to ensure that these are adequate and comply with the financial provisions indicated in the relevant TA documents, as well as ADB policies and guidelines.[36]

CTLA-TA staff receive around 100 queries daily from internal and external stakeholders, a significant portion of which are repetitive. This was seen as an indication of a significant demand for information and support as well as a need for a more efficient query handling system to unburden CTLA-TA staff so they could concentrate on more complex tasks like TA claims processing and advisory work.

About 25% of the TASU staff were new in 2020, and they had significant training needs on ADB's TA processes. Meanwhile, the lack of a centralized TA disbursement e-learning platform for clients meant there was no unified source for information or guidance, contributing to a high volume of repetitive inquiries. A temporary helpdesk was set up to only address inquiries specific to the TA claims portal. However, it was only open during working hours at ADB headquarters, leading to delays and frustration among clients seeking prompt assistance. Furthermore, without systematic tracking or monitoring of all client issues, CTLA-TA struggled to identify recurring problems and effectively address client needs.

ADB engaged Yellow.ai to develop a digital assistant capable of providing round-the-clock client support. The result was Technical Assistance Messaging Intelligence (TAMI), a chatbot deployed on various channels and set up to provide 24/7 virtual customer support to CTLA-TA's clients.

SUMMARY

What it is:
A digital assistant that provides real-time responses to disbursement-related queries.

Start of implementation:
2020

Implementation partners:
Controller's Department

Technology service provider:
Yellow.ai

In line with ADB's operational approach:
- Promoting digital development and innovative technologies

Work on the POC began in early 2020. The results led to the approval for the development of an MVP. Development and testing took place between October 2020 and March 2021 and involved key personnel from the Office of the Controller and the Information Technology Department.

The AI-powered chatbot can extract information from its knowledge database to provide relevant answers to users' queries. Natural language processing was used to allow the chatbot to interact with users in a conversational manner, as well as learn from every customer interaction.

TAMI was deployed on various channels and can provide 24/7 virtual customer support to CTLA-TA's clients. It also serves as a central repository and microlearning platform where users can search for

---

[36] ADB. 2020. *Technical Assistance Disbursement Handbook*. Manila.

and review TA policies and guidelines, as well as download forms and templates needed for claim processing purposes.

Users can either choose from the options automatically generated by TAMI or enter keywords to get answers to routine or procedural questions. They could also enable their microphones if they prefer to verbally ask their questions.

Users can provide their feedback at the end of each session after using TAMI to help further improve the digital tool. An integrated ticket management platform allows users to raise issues either directly through TAMI or via email. The platform streamlines the ticketing system by integrating Application Programming Interfaces for ticket assignment and facilitates automatic ticket transfers from processors to verifiers and approvers when agents are inactive. It also tracks the type and number of inquiries each agent handles and monitors their performance in resolving these issues. These features ensure the efficient handling of tickets.

TAMI is still live in 2024 and has gone through several enhancements to make it smarter and continuously improve user experience. It has garnered attention from over 5,000 customers, boasting a bot response accuracy rate exceeding 85% and a customer satisfaction score above 80%. TAMI has become an indispensable member of the CTLA-TA team, significantly enhancing productivity, efficiency, and the quality of customer service delivery.

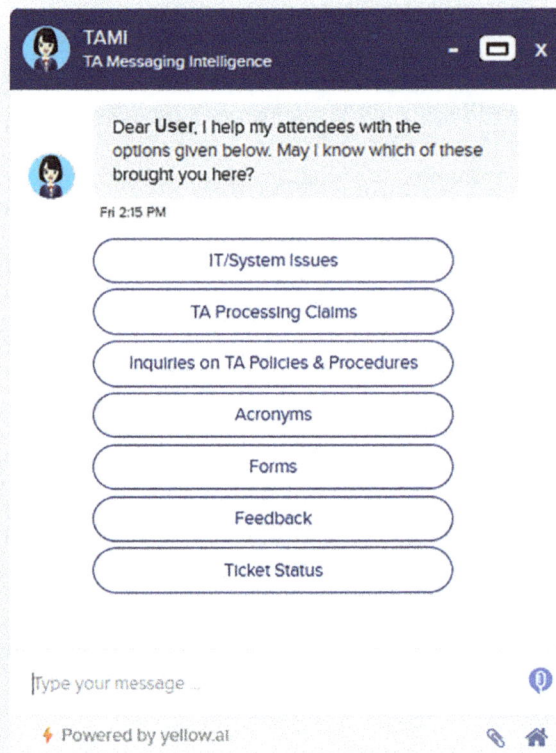

TAMI on the Technical Assistance Disbursement Handbook page on the ADB website.

MyADB Recruitment Intelligence

ADB has a manual recruitment process. About 100 applications are submitted for each open position, translating to an average of 45,000 submissions yearly. Recruiters spend several hours sorting, reviewing, and filing resumes.

Fatigue in reviewing, the lengthy hiring procedure, and slow response times can lead to ADB losing out on recruiting qualified candidates. In addition, ADB's process affects the line manager/recruiter experience.

ADB explored whether the offered platform of Impress.ai could address these issues. The platform was customized to fit ADB's requirements, with AI used to automate some parts of the recruitment sourcing process. Candidates uploaded their resumes to the digital platform, which parsed the documents, detected gaps, and prompted users to provide missing information.

Natural language generation was used to provide candidates with a realistic pre-screening interview experience. The AI-based platform asked situational judgment questions to candidates tailored to ADB's core values, as well as on position-specific skills and knowledge.

Meanwhile, machine learning enabled the platform to learn from its interactions with applicants, thus helping recruiters ensure that candidates were evaluated based on eligibility, competency, and technical proficiency. The platform also had a mechanism that allowed candidates to give feedback on their experience using the platform, which was dubbed MyADB Recruitment Intelligence (MARI).

In effect, the platform performed the tasks of line managers and recruiters by evaluating candidates' applications in relation to the requirements of the position they were applying for. Weights were given to the available responses so applications were scored quickly. The system recommended the disqualification of ineligible candidates. Shortlisted candidates were notified by the platform, which also scheduled their interviews.

SUMMARY

What it is:
MyADB Recruitment Intelligence was a digital assistant that was developed to automate some of the steps in ADB's recruitment process.

Start of implementation:
2019

Implementation partners:
Budget, People, and Management Systems Department

Technology service provider:
Impress.ai

In line with ADB's operational approach:
- Promoting digital development and innovative technologies

MARI was built in 2019. The platform was initially tested through an administrative assistant job opening in December 2019. Two additional tests—a national staff IT officer and a public management specialist—were conducted between February and March 2020 to train the bot to distinguish the differences in job openings. MARI showed that a digital assistant could screen applicants efficiently, saving time and money. Instead of manually evaluating applications, human resources personnel were able to focus their time on reviewing the qualifications of shortlisted candidates and conducting interviews.

They were also able to access the real-time updates and analytics on the dashboard, allowing them to identify bottlenecks in the hiring process to make the candidate workflow management smoother.

ADB

This year the market is not doing very well, and a lot of businesses are suffering losses or shutting down. One of your old clients is a victim of this market situation and has landed themselves in a crisis where they are running in losses. Your client is insisting that you help them to get a loan against their securities and that they are willing to mortgage their personal property. You have been their financial advisor for years now, and understand that the market is not in good shape and that they will fall into losses again if they take the risk of taking this loan. You are well aware that since the market has no scope of revival, this will be a bad decision. What will you do?

08 Mar, 02:30 PM

| You carefully draft a proposal of the current and projected status of the market, and present it to them to discourage them from applying for the loan | You send them various databases of market trends so that they can make an informed decision | You ask them to consult your manager and validate the market stability before taking the loan | You are transparent with them and firmly suggest that they do not go ahead with the mortgage |

Powered by *impress*

MARI was trained to conduct pre-screening interviews.
Image by Impress.ai

MARI improved candidates' recruitment experience and reduced the drop-off rate from 47% to 21.5%. The candidates also gave positive feedback on MARI. The FAQs module had a 95.4% accuracy rate in its responses to candidates. However, MARI was deprioritized due to larger human resource initiatives such as the Human Capital Management project and the New Operating Model.

Intelligent Integrity Due Diligence Platform

ADB recognizes the key role played by the private sector in reducing poverty and building climate resilience in Asia and the Pacific, which can help address development and infrastructure gaps, open trade opportunities, and generate gainful employment.[37] Thus, the bank aims to grow its private sector operations to support one-third of its total operations by 2024 as part of Strategy 2030.[38] However, ADB is careful in ensuring that this expansion does not come at the expense of due diligence. Through the Private Sector Operations Department (PSOD), it continues to conduct integrity due diligence (IDD) to adhere to high standards of corporate governance, integrity, and transparency. Stringent investigations identify real and possible risks posed by existing and potential clients. These measures are taken to ensure that the bank works with and provides funding to reputable partners who are neither involved in illegal, unethical, and disreputable activities nor work with individuals or partners who do so.

However, PSOD faces various challenges when doing IDD. The IDD process is time-consuming, requiring the manual completion of various checklists for each project. The annual IDD monitoring conducted to ensure risk per project is also completed manually. In addition, some documents may be in a language that reviewers may not be familiar with, thereby requiring more time for translation. These issues can endanger the reputation and operations of ADB and its partners, not to mention put the financial resources that ADB has been entrusted with at risk of being funneled to support illegal or unethical activities.

Through the Digital Learning Labs (formerly called the Digital Innovation Sandbox), a safe and neutral facility for ADB to test emerging technologies, ADB's Information Technology Department (ITD) and PSOD tested whether it was possible to use AI to conduct due diligence checks faster and avoid operational and reputational risks at the earliest possible instance. This would also have supported

SUMMARY

What it is:
Intelligent Integrity Due Diligence was a minimum viable product that used artificial intelligence to automate the integrity due diligence checklist preparation and investigative processes to make assessments faster and more efficient compared to manual reviews.

Start of implementation:
2019

Implementation partners:
Private Sector Operations Department

Technology service provider:
Neural Mechanics Inc.

In line with ADB's operational approaches:
- Promoting digital development and innovative technologies
- Expanding private sector operations

PSOD's goal of undergoing process change by improving its processes and internal efficiency, as highlighted in the Operational Plan for Private Sector Operations for 2019–2024.[39]

The MVP of Intelligent Integrity Due Diligence, which was developed with the support of Neural Mechanics Inc., focused on the IDD checklist. Specifically, it supported the IDD checklist preparation through Ultimate Beneficial Ownership searches, sanctions screening, adverse media searches, tagging, and footnote referencing to anchor search findings and generation of ownership diagrams where information

---

[37] ADB. 2021. *Technical Assistance for Supporting Integrity Due Diligence for Private Sector Infrastructure Projects in Central, West, and South Asia*. Manila.

[38] ADB. 2018. *Strategy 2030: Achieving a Prosperous, Inclusive, Resilient, and Sustainable Asia and the Pacific*. Manila.

[39] ADB. 2019. *Operational Plan for Private Sector Operations 2019–2024*. Manila.

is available online. The searches and findings were then presented and committed to the databases for easy retrieval and continuous monitoring. The news sentiment feature, meanwhile, assisted in filtering adverse information, enabling users to make faster assessments.

The product development stage aimed to integrate the generation of visual maps to show connections between individuals and organizations across multiple ADB projects. Search functionality was included to bring in results from different sources. The MVP was also designed to provide alerts, which were automatically categorized and prioritized as they came in.

Various lessons were learned in the development of the MVP. The insights that were generated from this exploratory initiative served as technical inputs into the development of the Integrity Intelligence System, an AI-driven platform that can scan multiple data sources for the screening and ongoing monitoring of sanctions, conducting assessments and investigations, and identifying integrity risk exposure for due diligence.

CHAPTER 4

ADB's Artificial Intelligence Blogs and Insights

Sketches of the different facets of AI.

Image by Rick Payne and team / Better Images of AI / Ai is... Banner / CC-BY 4.0

The previous sections have shown what ADB has done in the realm of AI. While this is already notable, especially considering that some were considered cutting-edge at the time of testing, a lot more can be done with the technology, which is still at its nascent stage of development.

AI can bring forth various innovations within and outside of ADB. It can potentially spur progress by unlocking previously unexplored solutions in the areas of climate change, health, education, and poverty reduction. However, the technology does have its share of risks, which is why it is essential to put into place policies and guidelines to regulate AI; otherwise, it may end up contributing to greater inequalities.

This section presents excerpts of thought pieces that were originally published on *Asian Development Blog* and *Development Asia* that tackled what AI could mean for global development. Links to the full entries are provided for readers who want to learn more about these topics.

AI and Development—Age of Abundance or Huge Divide?

Never has there been a major development shift without the advent of new technology and its adoption. From the internal combustion engine to the internet, technology is the driving force behind the evolution of societies.

Predictive and generative AI are quickly becoming the new drivers of development and economic growth. Some examples are as follows:

- **Transport.** ADB, in collaboration with ORIS, piloted a multiple-scenario digital analysis of road design in the upgrading of a section of the A380 highway in Uzbekistan. AI and digital simulations were used to analyze data on geography, construction materials, and pavement solutions to determine which design options were best suited for the local context and are more cost-efficient, sustainable, and climate-resilient.[40]
- **Climate change.** A Digital Urban Climate Twin is now being built in Singapore through the multidisciplinary project Cooling Singapore 2.0, which aims to improve the livability of the city-state, considering that is increasingly becoming warmer due to the urban heat island effect. Simulations are being run to determine what variations of strategies will best reduce heat and improve the city's resilience to climate change.[41]
- **Resettlement.** Through the Digital Innovation Sandbox, ADB tested whether AI could be used for monitoring and predicting resettlement outcomes in Mongolia. Anticipating potentially adverse outcomes is necessary to ensure that appropriate measures are in place to avoid or otherwise minimize, mitigate, and compensate for these impacts.

SUMMARY

Ozzeir Khan, director of Digital for Development Operations at ADB's Information Technology Department, writes about the potential of artificial intelligence to bring both abundance and division, and what developing economies can do to leverage the technology.

A version of this piece was published on *Asian Development Blog* (https://blogs.adb.org).

While AI is showing significant potential in driving development, the risk is real that generative AI could simultaneously usher in both greater prosperity and greater inequality.

Simply put, generative AI will spark the rise of new economies—and the fall of others.

Here are three actions governments could take to help slow the expansion of the AI divide in Asia and the Pacific.

Ensuring access to AI. Democratizing technology means that those who have not received formal training in information technology could still use generative AI to their advantage—provided they have an internet connection. However, 2.6 billion people do not.[42] The talent gap in AI in the People's Republic of China alone is expected to reach 4 million by 2030.[43]

---

40 ADB. 2023. *Building Greener, Resilient Transport Infrastructure: Innovating with Intelligence and Digital Twins in Road Design in Uzbekistan.* September. Manila.

41 Singapore Management University. 2022. *Cooling Singapore 2.0: A Step Towards Becoming a Climate Resilient and Regenerative City.* 8 September.

42 International Telecommunication Union. 2023. Population of Global Offline Continues Steady Decline to 2.6 Billion People in 2023. Press release. 12 September.

43 Nikkei. 2023. *The People's Republic of China has Set Off a Generative AI Boom, with a Talent Gap of 4 Million.* 25 September.

On a wider scale, governments and industries that are better prepared to adapt and harness the technology may find themselves emerging on top of the digital food chain, while those that cannot keep up with the fast pace will likely fall through the cracks, leading to greater inequality.

Imagine already prosperous economies accumulating even more wealth because they are in the position to heavily invest in generative AI and in upskilling their human resources. Contrast that with developing economies with less capacity to make both types of investments. Generative AI could also, directly or indirectly, lead to exploitation of human and other resources from developing countries.

To be part of the AI economy, it is vital for developing economies to integrate AI literacy into education, invest in AI research, and craft and pass relevant policies and regulations that promote responsible AI development and deployment.

Supporting policies for responsible use of artificial intelligence. Governments have a key role in helping shape the ethical use of AI. Thus, they need to take the lead in this new AI-driven era. They have the capacity to convene various stakeholders and adopt overarching strategies on the responsible and in setting up regulatory frameworks and guidance to ensure that these innovations are ethical, follow the "do no harm" principle, and support the improvement and protection of lives.

AI is evolving at a fast pace and requires constant changes in governance. In the coming years, AI will evolve from just text and multimedia generation that is now known for autonomous work and will need government support at a faster pace.

Putting basic artificial intelligence infrastructure in place. Understandably, developing economies face various financial constraints to invest in new technologies and build the necessary technological infrastructure. What they can do to compensate for this is to collaborate with partners, such as the private sector and multilateral development banks, to help bridge the gaps.

AI infrastructure is on top of the connectivity infrastructure investments that are driving digital growth in developing countries. This AI infrastructure focuses on data and models to drive value across sectors, leveraging foundation models like large language models. Without this basic AI infrastructure, the sectors in developing countries will not be able to reap the full benefits of developing innovation systems around AI to leapfrog connectivity gains compared to countries where these are in place.

While eyes are turned to the private sector to spearhead the development of AI infrastructure, it is important for governments to also be at the forefront, particularly in ensuring the responsible use of AI. It is paramount that developing countries are at an equal starting point for innovation discoveries and are part of the coming age of abundance rather than being part of the divide.

Generative Artificial Intelligence and Financial Inclusion

Generative AI is one of AI's most exciting and promising fields. It has captured the imagination of people worldwide as it can unleash the power of human creativity and imagination. For the finance sector, it can lower the barriers to entry for the unbanked and underbanked, allowing them to become part of the formal financial system and benefit from its services.

And expectations are high. According to a McKinsey report, the economic potential of generative AI for productivity and growth can be substantial, adding between $2.6 trillion and $4.4 trillion to world GDP and increasing labor productivity from 0.1% to 0.6% annually through 2040.

Continue reading at *https://blogs.adb.org/blog/your-questions-answered-generative-ai-and-financial-inclusion*.

AUTHORS

Lotte Schou-Zibell, a financial inclusion advisor for ADB, and ByeongJo (Jo) Kong, a digital technology specialist in data analytics and big data for ADB, answer questions about how generative artificial intelligence can aid in the effort to expand financial inclusion in Asia and the Pacific.

The original version of the blog was published on 11 December 2023.

Could ChatGPT Be a Game Changer for International Development?

It seems that everyone is talking about ChatGPT and its AI "colleagues" and how they will change the way we work, including those of us who work in development. Some people say it will help us to solve the big global challenges—for example, inclusive and sustainable development. Others are concerned that it could have still undetermined negative impacts.

To help us find the answer, we decided to ask ChatGPT directly.

Continue reading at *https://blogs.adb.org/blog/could-chatgpt-be-game-changer-international-development-we-asked-it.*

AUTHORS

Cass R. Sunstein, Robert Walmsley University Professor at Harvard, and Susann Roth, advisor, Department of Communications and Knowledge Management, share their thoughts on the potential and limitations of large language model artificial intelligence such as ChatGPT and Bard in the world of international development.

This piece was originally published on 11 April 2023.

Six Ways to Prepare Teachers in Asia and the Pacific for the ChatGPT Era

The lack of qualified teachers is becoming the top hurdle in developing countries as they upgrade their education systems to prepare for the needs of 21st-century students.

Many low-income countries and some middle-income countries in Southern Asia need an additional seven million teachers by 2030, including 1.7 million in primary and 5.3 million in secondary education, according to UNESCO.

Continue reading at *https://blogs.adb.org/blog/six-ways-prepare-teachers-asia-and-pacific-chatgpt-era.*

AUTHORS

The region needs more and better educated teachers, particularly in the area of artificial intelligence in order to meet the needs of students in the years ahead, according to Jeffrey Jian Xu, senior education specialist, ADB Human and Social Development Sector Office, Sectors Group; and Jukka Tulivuori, social sector specialist, ADB Human and Social Development Sector Office, Sectors Group.

The original piece was published on 26 July 2023.

Five Ways to Reap the Benefits of the Technological Revolution

Technologies including AI, big data, cloud technology and Internet of Things, are fundamentally changing the way we work, live, and relate to one another. The shifts and disruptions they are causing are so massive that the stakes for success are as high as the perils.

There are currently about 20 billion devices connected to the internet, and almost half of enterprises globally are adopting cloud technology as an operative norm. In 2020, it was predicted that by 2022, one in five workers engaged in non-routine tasks will rely on AI. AI augmentation will help business enterprises to recover 6.2 billion hours of worker productivity.

Continue reading at *https://blogs.adb.org/blog/five-ways-reap-benefits-technological-revolution.*

AUTHORS

Governments and development partners need to take a methodical approach to the adoption of artificial intelligence technologies in collaboration with the private sector, says Arun Ramamurthy, principal infrastructure specialist (Digital Technology), ADB Transport Sector Office, Sectors Group.

The original piece was published on 9 November 2020. Minor revisions were made on this excerpt.

Using Machine Learning on Satellite Images to Map Poverty

The Sustainable Development Goals (SDGs) comprise 169 specific, time-bound, and measurable targets for 2030 to leave no one behind. The SDGs carry forward and extend global efforts to achieve the Millennium Development Goals for attaining socioeconomic development, with disaggregation by different dimensions, such as age, sex, geographic areas, and income. Collecting and compiling these indicators translate to enormous work for national statistical systems across different nations.

Moreover, financial resources from bilateral grants and multi-donor trust funds for supporting statistical programs are limited and sparse. National governments may also not have enough budget to finance statistical development programs for various reasons. In view of scarce resources, many national statistical systems resort to alternative methods to meet the growing demand for SDG data along with other emerging data requirements for development planning.

AUTHORS

This article presents a study that examined the feasibility of applying computer vision techniques to satellite data of the Philippines and Thailand to produce poverty maps.

This piece was written by Arturo Martinez, Jr., statistician, Economic Research and Development Impact Department at ADB.

The original piece was published on 11 September 2020.

Continue reading at *https://development.asia/insight/ using-machine-learning-satellite-images-map-poverty.*

Artificial Intelligence and Machine Learning in the Time of COVID-19

The COVID-19 pandemic has caused us major setbacks, but artificial intelligence and machine learning can help fill some of these gaps. By learning from AI's successes and consequences, we can develop responsible, data-driven solutions to help us through the current crisis.

What is the potential of AI and machine learning? AI refers to a number of ways machines are built to perform with human-like intelligence. Under its broad field is a subset called machine learning, which involves the processing of massive amounts of data combined with sophisticated algorithms into machines so they can "learn" specific tasks automatically. This helps us pinpoint patterns or make decisions from a sea of information.

Continue reading at *https://blogs.adb.org/blog/artificial- intelligence-and-machine-learning-time-covid-19.*

AUTHORS

Researchers are finding new uses for artificial intelligence and machine learning during the pandemic, but they are not a silver bullet.

The piece was written by Bruno Carrasco, director general, Climate Change and Sustainable Development Department, ADB; Stephanie Sy, chief executive officer and lead data scientist, Thinking Machines; and Woochong Um, managing director general, Office of the President, ADB.

This was originally published on 29 July 2020.

Data and the Artificial Intelligence Gold Rush: Who Will Win?

AI will someday know you better than you know yourself. That day may be sooner than we realize with the amount of data collected on all humans and their environments increasing exponentially. So where are the rules, and what are our rights?

Over the past few centuries, data has been collected at high levels: primarily on companies, countries, societies, cultures, religions, and other high-level aggregations. With the data age in full swing, we are delving into the frontier of individual data—a level previously unreached in terms of deeply knowing and connecting humans.

Continue reading at *https://blogs.adb.org/blog/data-and-artificial-intelligence-gold-rush-who-will-win*

AUTHORS

Ozzeir Khan, director of Digital for Development Operations at ADB's Information Technology Department, argues that the exponential growth of data and artificial intelligence is creating a tug-of-war between data for profit and data for the common good. He points out the fundamentality of protecting basic human data rights in this struggle.

The original piece was published on 28 January 2020.

www.ingramcontent.com/pod-product-compliance
Lightning Source LLC
Chambersburg PA
CBHW050055220326
41599CB00045B/7415